Brilliant Support Activities

Understanding Physical Processes

Roy Purnell, Janet O'Neill and Alan Jones

Brilliant Publications

We hope you and your class enjoy using this book. Other books in the series include:

Science titles

Understanding Living Things	978 1 897675 59 5
Understanding Materials	978 1 897675 60 1

Language titles

Sentence Level Work	978 1 897675 33 5
Text Level Work	978 1 903853 00 9
Word Level Work – Phonics	978 1 897675 32 8
Word Level Work – Vocabulary	978 1 903853 07 8

Published by Brilliant Publications
website: www.brilliantpublications.co.uk

Sales
BEBC (Brilliant Publications)
Albion Close, Parkstone, Poole, Dorset BH12 3LL, UK
Tel: 01202 712910   Fax: 0845 1309300
email: brilliant@bebc.co.uk

Editorial
Brilliant Publications
Unit 10, Sparrow Hall Farm, Edlesborough, Dunstable LU6 2ES, UK
Tel: 01525 222292   Fax: 01525 222720

The name Brilliant Publications and its logo are registered trademarks

Written by Roy Purnell, Janet O'Neill and Alan Jones

Designed and illustrated by Small World Design

The authors are grateful to the staff and pupils of
Gellideg Junior School, Merthyr Tydfil for their help.

Printed in the UK
First Published in 2000, reprinted 2002, 2008.
10 9 8 7 6 5 4 3
ISBN 978 1 897675 61 8

Contents

Introduction to the series

This series is designed to help slower learners or pupils with learning difficulties at Key Stage 1 and 2 develop the essential skills of observation, predicting, recording and drawing conclusions. These pupils often have been neglected in more conventional commercial schemes of work. The books contain a mixture of paper-based tasks and 'hands on' activities. Symbols have been used to indicate different types of activities:

 What to do

 Think and do

 Read

 Investigate

The sheets support the attainment targets for Key Stage 2 Science, Section 1: Scientific enquiry and Section 4: Physical processes. The practical investigations use materials readily available in most primary schools. The activities have been vetted for safety, but, as with any classroom-based activity, it is the responsibility of the class teacher to do a risk assessment with her/his own pupils in mind.

The sheets usually introduce one concept or National Curriculum statement per sheet (unless they are review sheets). The sheets are designed for use by individual pupils or as a class activity if all the class are working within the same ability range. They can be used in any order, so that you can choose the sheet that best matches a pupil's needs at that particular time. As with any published activities, the sheets can be modified for use by specific pupils or groups. The sheets can be used as a support for your present schemes, as an assessment task, or even as a homework task. If used for assessment purposes, then you will need to devise a marking scheme or level indicator. Generally the sheets are designed for use with levels 1–3 but some can be used at level 4.

The sheets use simple language and clear, black line illustrations to make them easy to read and understand. They have been tested to check that they can be understood by pupils with learning difficulties. Although the sheets have a reduced vocabulary, they encourage pupils to produce written responses and to develop their writing skills.

No particular reference has been made to any type of disability as the activities should be accessible to a wide range of pupils and it is up to the teacher to select the most suitable modes of access to match the needs of her/his pupils. For example, the activities could be photo-enlarged, converted to raised, tactile diagrams, or recorded on an audio tape.

Introduction to the book

The topics in this book help pupils understand physical processes. They reinforce methods of scientific enquiry by requiring pupils to plan and carry out practical activities, consider evidence, and present ideas and conclusions. The sheets focus on forces and motion, electricity, light and sound, as well as on the Sun and Moon and their relationship to the Earth.

The worksheets in this book overlap and you will find that several statements of the National Curriculum are covered several times in a number of different ways. This is to allow you to use the worksheets to repeat work on particular concepts to reinforce your pupils' learning. However, the worksheets are not designed to be used in any particular sequence. They are not a teaching scheme, but are a resource which you can use to enrich or augment your own particular scheme of work according to the needs of your pupils.

Some worksheets encourage an open-ended response, others are designed to lead pupils to a particular answer. Some start with easy tasks and progress to more difficult extension activities which we have called 'Think and do'. Others are at one level of difficulty. The variety is designed to give the worksheets flexibility and to allow you to select the most appropriate worksheet for your pupils.

Stretch and shape

Read

Forces can make things change shape.

What to do

Draw or write what happens. Put a tick ✓ to show if it is a **stretch** or a **squash**.

	Draw here	Stretch	Squash
Hammer / Plasticine			
Exercise chest expander			
Soft cushion			
Bungee jumper			

Think and do

Look at this toy plane.

Propeller

Elastic band

What happens to the elastic band when the propeller is wound up?

. . . .

What happens when you let go of the propeller? . . . .

The strong force of gravity

Read

Gravity is a force. It pulls things towards the ground.

What to do

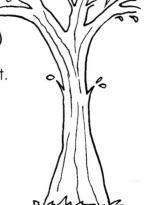

Two apples fall off the tree at the same time from the same height. One apple is big and one is small.

Which will hit the ground first?. . . . .

When an apple falls off a tree, it is pulled down by the

force of **g** . . . .

A man named Galileo dropped a big iron ball and a small iron ball from the top of a tall tower in Pisa, Italy.

Which one do you think hit the ground first?

. . . .

You can investigate this.

Use two marbles of different sizes and a metal tray. Listen to the 'clonk' as they hit the tray.

Think and do

Will the swing times be the same for both balls of Plasticine? They have the same length of string.

Time for 20 swings.

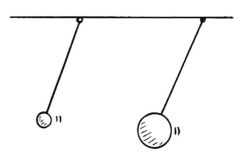

Plasticine

Air resistance

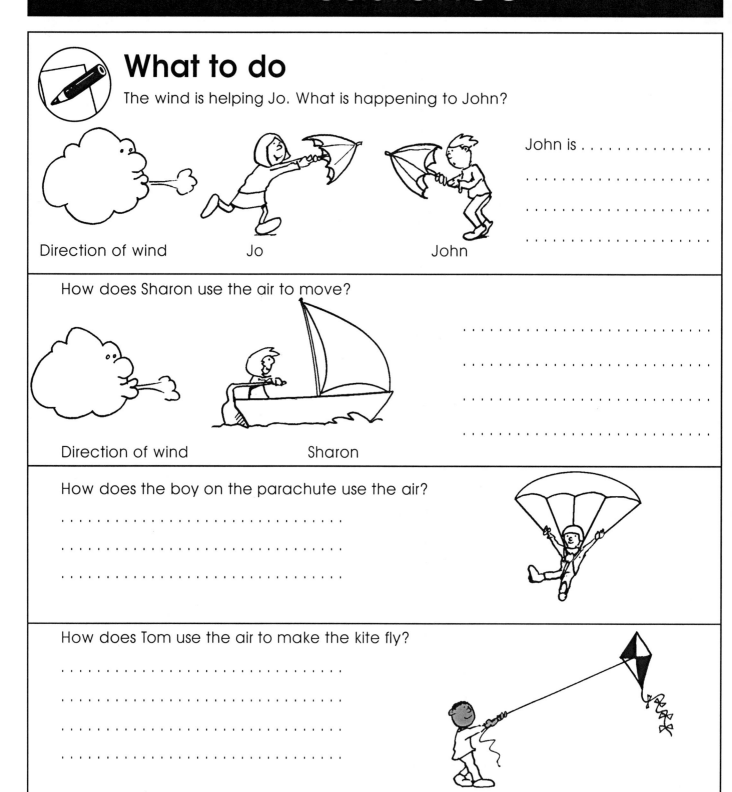

What to do

The wind is helping Jo. What is happening to John?

Direction of wind   Jo   John

John is . . . .

. . . .

. . . .

. . . .

How does Sharon use the air to move?

Direction of wind   Sharon

. . . .

. . . .

. . . .

. . . .

How does the boy on the parachute use the air?

. . . .

. . . .

. . . .

How does Tom use the air to make the kite fly?

. . . .

. . . .

. . . .

. . . .

Think and do

How does a dandelion seed make use of the air?

How does the shape of the seed help?

What to do

List some **forces** in action in each of the pictures.

. . . .

. . . .

. . . .

. . . .

. . . .

. . . .

Think and do

What force acts to try to stop things moving?

f _ _ _ _ _ _ _ _

You can cut up the letters and rearrange them.

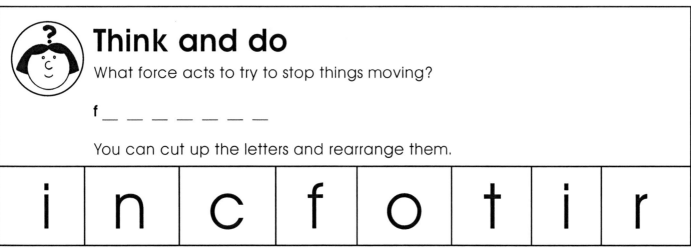

i	n	c	f	o	t	i	r

Balancing forces

Read

A seesaw balances when the force on one side equals the force on the other side.

Investigate

Make a seesaw with a ruler and pencil. Make it balance by moving the coins.

Draw where the coins have to be to make the seesaw balance.

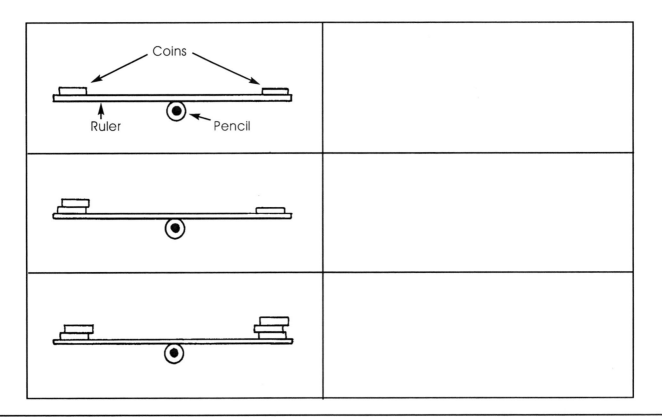

Coins

Ruler   Pencil

Think and do

An adult who is twice the mass of a child uses a seesaw.

How could they make the seesaw balance?

Draw what happens now.

What to do

What is the name of the force Becky is using to move the trolley?

. . . .

What is the name of the force Becky is using to stop the trolley from moving?

. . . .

The trolley is now full of shopping.
What do you think will happen if Becky lets go?

. . . .
. . . .

Think and do

Draw or write what will happen when Sam helps Becky. Show the direction of pushes and pulls.

Rough stuff

Read

A moving car runs out of petrol and soon stops. The stopping force is called **friction** between the wheels and the road.

Investigate

How far does the car go?

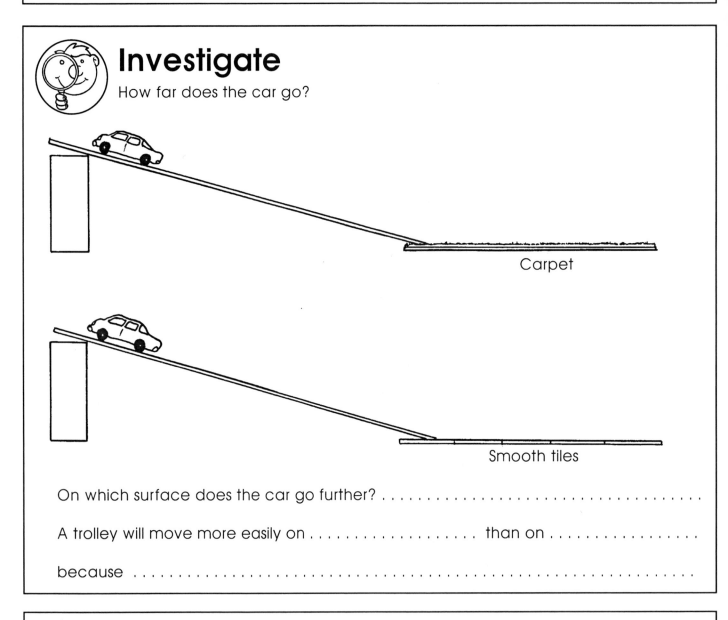

Carpet

Smooth tiles

On which surface does the car go further? . . . .

A trolley will move more easily on . . . . than on . . . .

because . . . .

Think and do

Give two reasons why roads are flat but slightly rough.

1. . . . .

2. . . . .

What happens to moving cars when there is ice on the road?

. . . .

Bungee jumping

Read

Fred is a bungee jumper.
He is fixed by a thick elastic rope
to the top of a very high crane.

What to do

When Fred jumps, why is he not hurt?. . . . .

Why is the rope made of elastic and not ordinary rope?

. . . .

Use these words to complete the sentences:

stretches	pulls	slows	gravity	longer

1. The rope **s** . . . . and gets **l** . . . . when Fred jumps.

2. When the rope stretches, it **s** . . . . down Fred's fall.

3. **G** . . . . pulls Fred downwards.

4. The stretch of the rope **p** . . . . Fred upwards.

Think and do

Savita is twice as heavy as Fred.
What happens if Savita uses the
same bungee rope?

Bigger masses on the rope will stretch

the elastic rope **more** ☐   **less** ☐

Tick the correct box.

Moving things

What to do

Write **push** or **pull** to show what is happening in each picture.

How does James put on
his socks? . . . .

How does Sally move
the piano? . . . .

How does Marie move
the buggy? . . . .

How does Jack move the
zip down? . . . .

How does the horse move
the wagon? . . . .

How does the man move
the car? . . . .

Think and do

Push or pull? Write the correct word.

1. How do we open a cupboard door?. . . . .

2. How do we put on trousers?   . . . .

3. Rugby players in a scrum.   . . . .

What do forces do?

Read

You cannot see a force, but you can see what a force does.

What to do

Forces are being used in these pictures.
Show the direction of the forces.

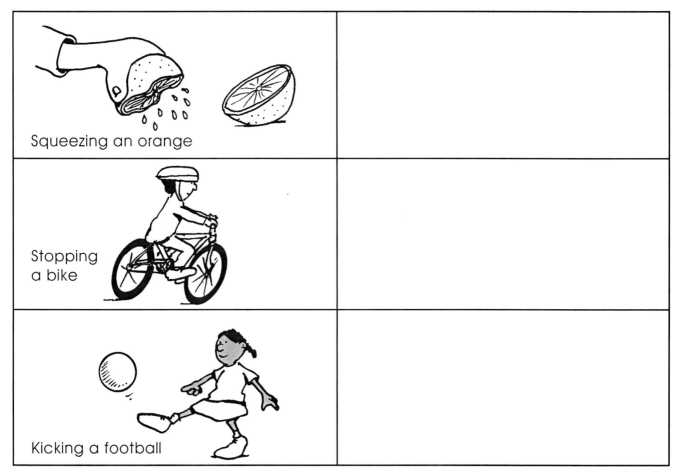 Squeezing an orange	
Stopping a bike	
Kicking a football	

Think and do

What force makes the child slide down the slide?

. . . .

How does the child stop at the bottom?

. . . .

Forces have direction

 ## Read

The force of the wind can move a sailing boat forward.
The arrow shows the direction of the force.

Force of wind

 ## What to do

Draw arrows on the pictures to show the direction of the forces.

Man pushing buggy

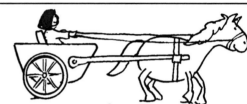

Horse pulling cart

Girl moving sledge

Woman pushing car

Boy pulling fish from water

Wind blowing tree

Bird pulling worm from the ground

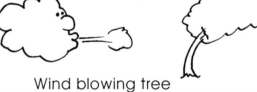

Dog pulling tablecloth

 ## Think and do

Mary lifts her heavy school bag off the floor.

Draw a picture to show what she does.

Draw arrows to show the direction of the force.

Equal and opposite

 ## Read

Two tug of war teams pull in opposite directions.

Team A   Team B

When the two forces are the same, the teams do not move.
Team A is working against Team B.

 ## What to do

Team A gets an extra person.
Write or draw what happens.

Use these words to complete the sentences.

same   **opposite**

Forces can help each other if they are going in the . . . . direction.

Forces can balance each other if they are going in the . . . . direction.

 ## Think and do

Why is it easier to pick up a heavy
bucket with two hands than with
one hand?

. . . .

. . . .

What to do

Look at the pictures of muscles working.
Draw arrows to show the direction of the force by the muscles.

Think and do

Think of another example. Draw it.

Magnets

 ## Read

A magnet can pick up iron paper clips.
When it does this, the attractive force of
the magnet is stronger than the force of gravity.

 ## Investigate

There are several shapes of magnet:

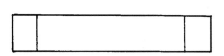

Bar

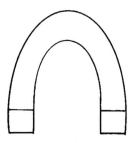

Horse shoe

Ring magnet

One end is called the **North pole**.

The other end is called the **South pole**

Is the attractive force the same at both ends?

Use iron paper clips to investigate this.

 ## Think and do

Where are magnets used in the home?

· ·

· ·

· ·

· ·

· ·

Magnets - push or pull?

Investigate

Use some magnets to try this.

First predict if they will **repel** (push away) or **attract** (pull together).

	My prediction	What happened?
Put two different ends of a magnet close to each other.		
Put the same ends of two magnets together.		
Put one magnet on top of the other, matching the colours.		

What to do

Use **North** or **South** to complete the sentences:

1. A . . . . pole attracts a South pole.
2. A North pole repels a . . . . pole.
3. A South pole repels a . . . . pole.

Think and do

A compass is a magnetic pointer on a pin.
The pointer can move about freely.
Draw the pointer on the compass shape.

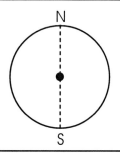

Force words

What to do

Put a tick ✓ by the words that we use when describing a **force**.

gravity ☐   pull ☐   sheep ☐   zebra ☐

push ☐   radio ☐   squeeze ☐   mat ☐

change shape ☐   food ☐   start ☐   spring ☐

stopping ☐   window ☐   apple ☐   braking ☐

change direction ☐   moving ☐   elastic ☐

stretch ☐   orange ☐   friction ☐

classroom ☐   spring balance ☐   paper ☐

shove ☐   pencil ☐   speed ☐   lifting ☐

Useful electricity

 ## Read

Electricity comes from batteries or power stations. We call electricity from power stations **mains electricity.**

 # What to do

Use these words to complete the sentences.

bulb

kettle

washing machine

calculator

cat

television

radio

1. A **r** . . . .   needs electricity.
2. A **b** . . . .   uses electricity to make light.
3. A **t** . . . .   needs electricity to give a picture.
4. A **w** . . . .   **m** . . . .   uses electricity to get clothes clean.
5. A **k** . . . .   boils water using electricity.
6. A **c** . . . .   gets its energy from food.
7. A **c** . . . .   uses electricity to do sums.

 # Think and do

Find one new thing that uses a battery for electricity.
Find one new thing that uses mains electricity.

Write a sentence about each one on the back of this sheet.

Will the bulb light?

Read

For a bulb to light you **must** have a complete circuit.
The electricity must go from the battery, through the wire, through the bulb and back to the battery.

What to do

Trace the path of the electricity with your finger.
You can investigate this if you wish.

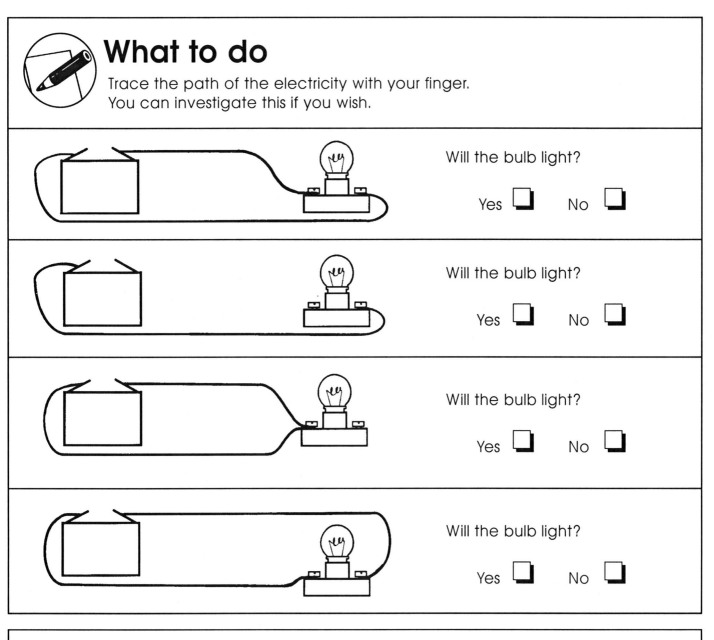

Will the bulb light?

Yes ☐   No ☐

Will the bulb light?

Yes ☐   No ☐

Will the bulb light?

Yes ☐   No ☐

Will the bulb light?

Yes ☐   No ☐

Think and do

Draw in the wires to
make the bulb light.

 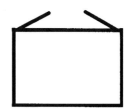

Try these circuits

What to do

Look at the pictures. Predict if the bulbs will light.
Make the circuits and see if you were correct.

	My prediction	What happened?
	On/Off	On/Off
	On/Off	On/Off
	On/Off	On/Off
	On/Off	On/Off
	On/Off	On/Off

Think and do

Draw a circuit with three bulbs. Make sure all three bulbs will light.

Electricity pass or not?

Read

Electricity passes through metal.
It does not pass easily through plastic, wood, brick or concrete.

What to do

Look at the pictures. Predict which things
will let electricity pass through. Make a circuit.
Add each thing to the circuit in the gap and see if you were correct.

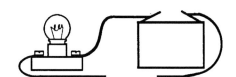

	My prediction The bulb will be ...	What happened? The bulb is ...
Iron nail	On/Off	On/Off
Plastic spoon	On/Off	On/Off
Metal spoon	On/Off	On/Off
Wood	On/Off	On/Off
Coin	On/Off	On/Off
Cotton wool	On/Off	On/Off

Think and do

Things that let electricity pass through are called **conductors.**

Why are metal electric wires covered with plastic?

. . . .

How to dim bulbs

Read

Batteries push electricity around circuits through wires and bulbs.
It is harder to push electricity through bulbs than through wires.

One bulb   Two bulbs

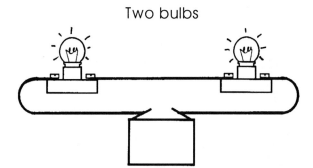

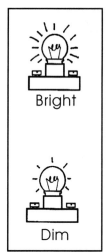

Bright

Dim

It is harder to push electricity through very long wires than through short wires.

Short wire   Very long wire

What to do

1. In which circuit will the bulb be **brightest** A, B or C?

Answer

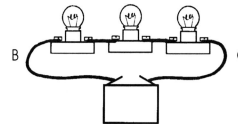

2. In which circuit will the bulb be **dimmest** A or B?

Answer

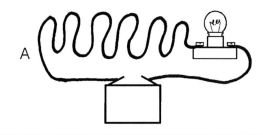   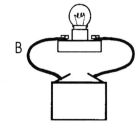

Dim and bright bulbs

Read

Batteries push electricity around a circuit.
When two batteries are joined in a circuit,
the push is twice as strong. Then the bulb is brighter.

Bright

Dim

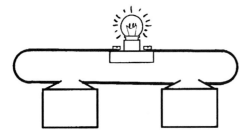

Two batteries and two bulbs will have the same brightness as one bulb and one battery.

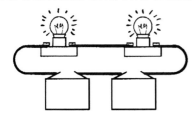

Three bulbs and three batteries will be brighter than two batteries and three bulbs.

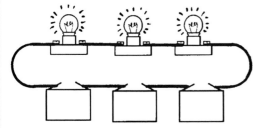

What to do

Draw a circuit with two bulbs and three batteries.

Look at these circuits:

In which one are the bulbs the brightest? . . . .

In which one are the bulbs the least bright? . . . .

A
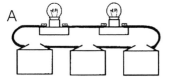

B
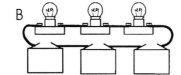

C
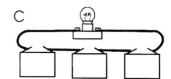

Warning: If you use too many batteries the bulbs can blow. The electricity will melt the wire in the bulbs.

Switches

What to do

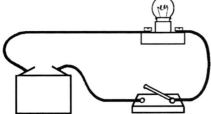

What happens to the bulb when the switch is closed?

. . . .

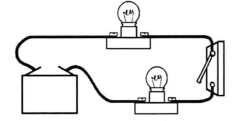

What happens to the bulbs when the switch is closed?

. . . .

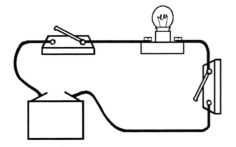

What happens when one switch is closed?

. . . .

What happens when both switches are closed?

. . . .

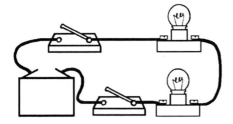

What happens when one switch is closed?

. . . .

What happens when both switches are closed?

. . . .

You can investigate this.

Think and do

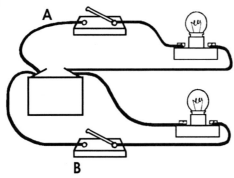

What happens when switch **A** is closed?

. . . .

What happens when both switches are closed?

. . . .

Read

Scientists draw plans of circuits.
They are called **circuit diagrams.**
Each thing in a circuit has a different symbol.

Picture	=	Name	=	Symbol
	=	wire	=	
	=	battery	=	
	=	bulb	=	
	=	switch	=	

A battery joined to a bulb would have a circuit diagram like this:

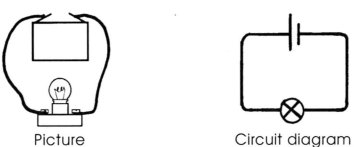

Picture   Circuit diagram

What to do

Draw circuit diagrams for these circuits.

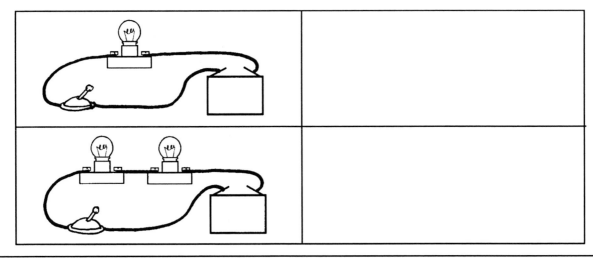

More circuit diagrams

Read

In **series** means joined in a line.
These two batteries are connected in series.

We can draw diagrams to show electric things connected in series.

Picture		**Diagram**

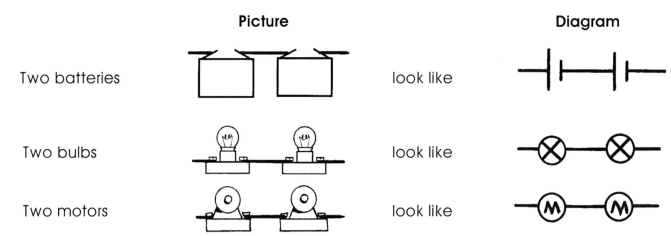

Two batteries — look like

Two bulbs — look like

Two motors — look like

A circuit diagram for a circuit with two batteries and two bulbs joined in series would look like this:

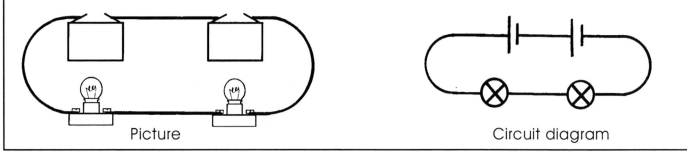

Picture   Circuit diagram

What to do

1. Draw a circuit diagram of one battery and two motors joined in series.

2. Draw a circuit diagram of two batteries, two bulbs and a switch joined in series.

Electric symbols and pictures

What to do

Draw lines to join the words to the pictures and symbols. One has been done for you.

Pictures	Words	Symbols

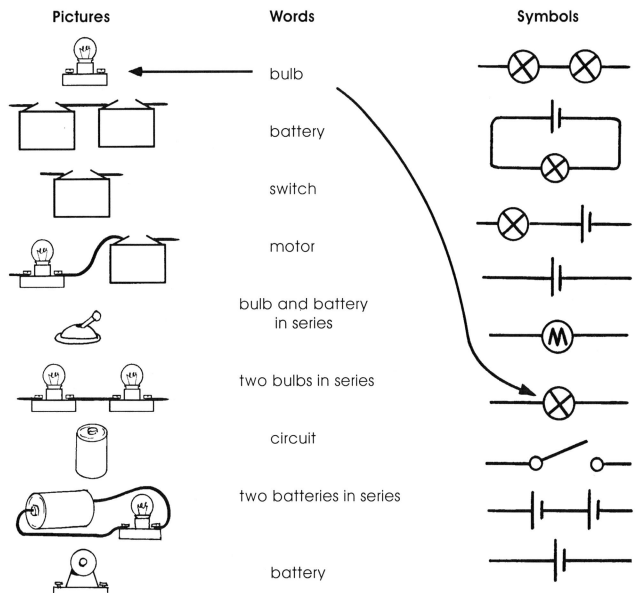

bulb

battery

switch

motor

bulb and battery
in series

two bulbs in series

circuit

two batteries in series

battery

Think and do

Draw a circuit diagram with one motor, one switch and one battery.

Electric words

What to do

Put a tick ✓ by the things that need electricity to work.

sheep ☐   bulb ☐   tree ☐   Christmas lights ☐

flower ☐   wire ☐   cat ☐

motor ☐   lollipop ☐   television ☐   wall ☐

switch ☐   bread ☐   radio ☐   road ☐

doorbell ☐   door ☐   refrigerator ☐

torch ☐   computer ☐   street light ☐

cable ☐   headlight ☐   dog ☐   circuit ☐

vacuum cleaner ☐   coal fire ☐   underground train ☐

battery ☐   traffic lights ☐   book ☐

Read

You can see something that gives out light, like a candle. The candle is called a **light**.

You can see anything that light bounces off, like a tree.

You cannot see when it is completely dark.

 ← Mask

What to do

Draw light lines to show why Sam can see Jamie.

Sam

Jamie

Think and do

Draw lines to show how Sam can see the light even when he's got his back to it.

 light

Sam

mirrror

Light makers

What to do

Put a tick ✓ by the things that make light.

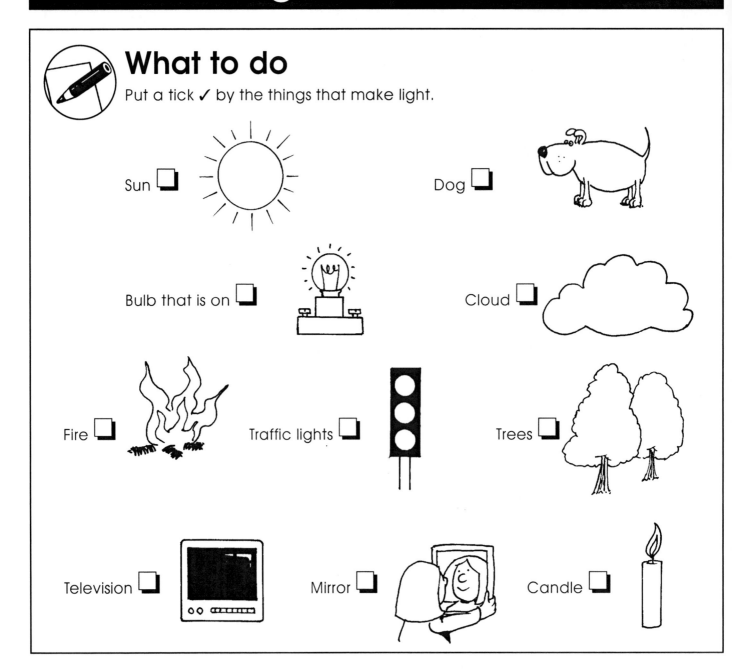

Sun ☐

Dog ☐

Bulb that is on ☐

Cloud ☐

Fire ☐

Traffic lights ☐

Trees ☐

Television ☐

Mirror ☐

Candle ☐

Think and do

Make a list of all the different ways you can think of to light up a dark room.

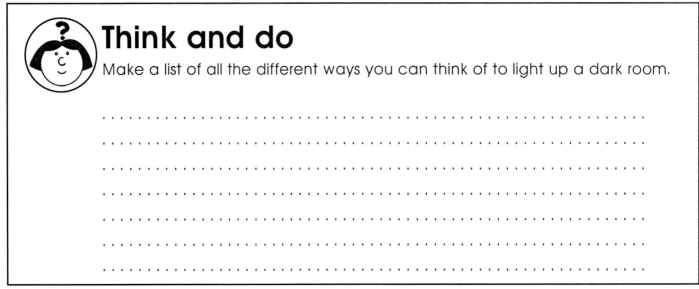

. . . .

. . . .

. . . .

. . . .

. . . .

. . . .

. . . .

Shadows

Read

A shadow is made when light does not pass through an object.

A pencil will make a shadow

Clear plastic or glass will not make a shadow. The light passes through.

What to do

Draw the shadow of the piece of card.

Card

Draw the shadow of the tree.

The shadow looks like the shape of the thing.
The shadow is made because the light is blocked off.

Think and do

Why does the shadow of a cat
have the same shape as the cat?

. . . .

Draw a picture to show this.

Mirrors reflect light

Read

We see the pencil because the mirror reflects light back into our eyes.

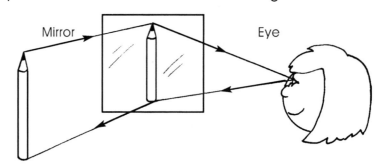

What to do

Draw where the mirror needs to be for you to see around a corner.

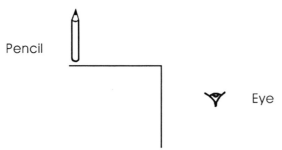

Draw two mirrors so that you can see the pencil.

The pencil is behind you. Where does the mirror need to go?

Light words

What to do

Find the words that we use when we talk about light.

The words are listed underneath the grid.

z	s	h	i	n	e	t	r	e	d
o	q	i	n	d	i	g	o	r	s
r	j	r	e	f	l	e	c	t	s
a	x	y	z	g	h	j	z	w	h
n	y	e	l	l	o	w	d	b	a
g	v	w	w	x	j	d	g	l	d
e	h	n	m	p	q	r	z	u	o
z	v	i	o	l	e	t	z	e	w
g	r	e	e	n	q	r	w	r	s
e	y	e	z	m	i	r	r	o	r

shine	eye
shadow	red
reflect	orange
mirror	green
yellow	blue
indigo	violet

Making sounds

What to do

How can these things make sounds?
Use these words:

pluck	shake	tap
beat	blow	bang

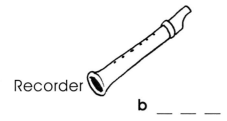

Recorder

b _ _ _

Cymbals

b _ _ _

Maracas

s _ _ _ _

Triangle

t _ _

Guitar

p _ _ _ _

Drum

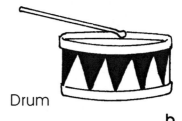

b _ _ _

Think and do

1. To make a louder sound with the drum you bang the drum
   h _ _ _ _ _.

2. To make a louder sound with the recorder you have to
   b _ _ _ harder.

Good vibrations

What to do

Put some grains of rice on the skin of a drum.
Give the drum a few bangs.
Draw what happens to the rice.

Did you notice the skin of the drum moving? Yes ☐ No ☐

The drum skin **vibrates** up and down.

Investigate

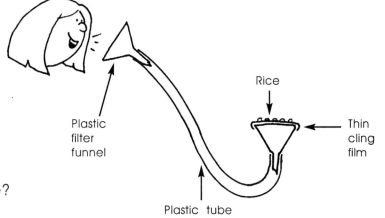

Rice

Plastic
filter
funnel

Thin
cling
film

Plastic tube

Make this:

Speak into the open end.

What happens to the grains of rice?

. . . .

What has caused the rice to move?

. . . .

Could you see the plastic **vibrating?** Yes ☐ No ☐

This works a bit like a microphone.

Think and do

How can you alter the note a drum makes?

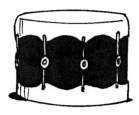

What to do

A radio is playing loud music in the room next to your classroom.

You can hear the music in your room.

Use these words to complete the sentences:

air	wood	glass	window

The sound passes easily through the **a** . . . . in the rooms.

The sound passes less easily through the **g** . . . . in the windows.

The sound passes through the **w** . . . . frame.

The sound passes through the **w** . . . . in the floors.

Think and do

Turn a radio on quietly. Put it in a cardboard box.

Can you still hear it?   Yes ☐   No ☐

Surround the radio with different materials.

Find the best one to muffle the sound.

The best material to muffle the sound is . . . .

What to do

Three elastic bands are stretched and twanged. Which has the highest pitch? Tick the correct box.

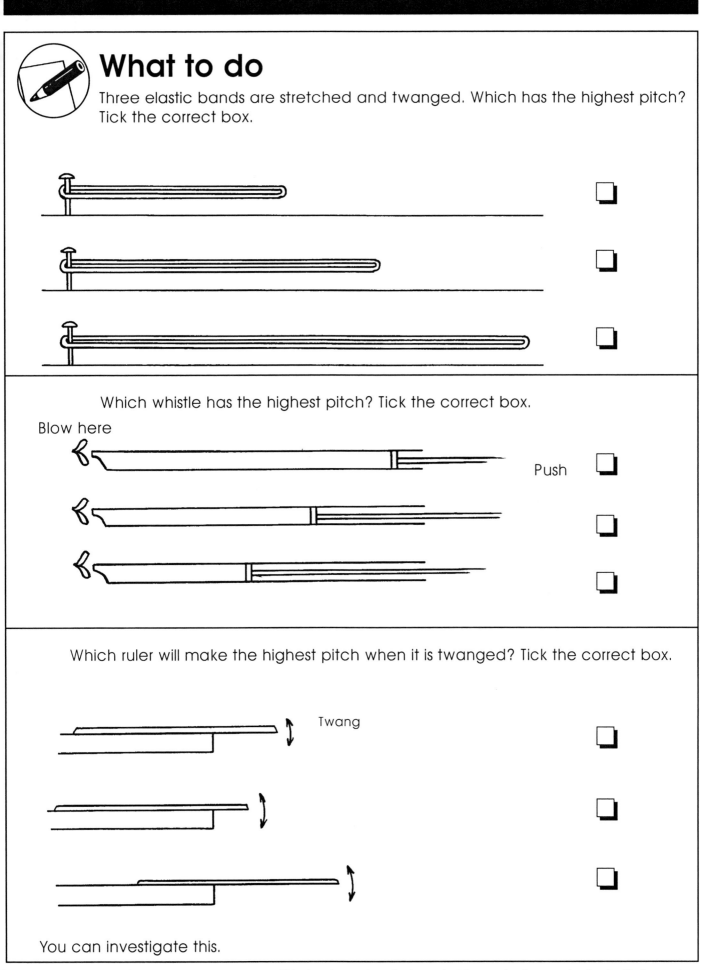

Which whistle has the highest pitch? Tick the correct box.

Blow here

Push

Which ruler will make the highest pitch when it is twanged? Tick the correct box.

Twang

You can investigate this.

This sheet may be photocopied for use by the purchasing institution only.

What to do

Rearrange the letters to make words we use when we talk about sound.

nesio _____   durm _____

olinvi _____   tremput _____

raguit _____   ludo _____

raiod _____   troomben _____

sotf _____   usicm _____

samarac _____   are _____

tlriaeng _____   scybmal _____

decorerr _____   piona _____

The Earth, Sun and Moon

 ## What to do

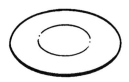 A plate   An orange   A box

The Earth, Moon and Sun are all the same shape as . . . .

Which is which? Put a tick by the correct word.

The Earth moves around the   Sun ☐   Moon ☐

The Moon moves around the   Sun ☐   Earth ☐

Planets, like Mars, move around the   Earth ☐   Moon ☐   Sun ☐

Label the diagram to show the Sun, Moon and Earth.

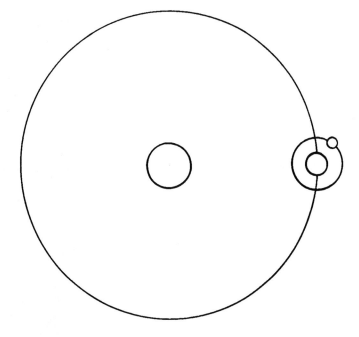

 ## Think and do
Put the planets Venus, Mercury and Mars on the diagram above.

The turning Earth

Read

We live on a planet called Earth.
Look at the picture to see where we live.
The Earth spins once in 24 hours (or 1 day).
In half a day the Earth turns half the way round.

How many hours does this take? . . . .

The United Kingdom is here

What to do

When we are facing the Sun, it is daytime.
When we are facing away from the Sun, it is night-time.

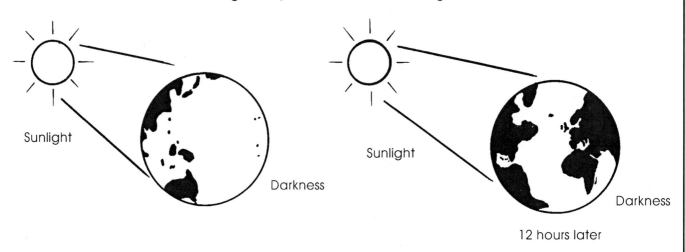

Sunlight

Darkness

Sunlight

Darkness

12 hours later

Shade the parts of the Earth that will be in darkness in each picture.

Think and do

Explain why it is cold at night.

. . . .

. . . .

. . . .

What happens to day and night at the North and South poles?

. . . .

. . . .

The Sun seems to move

 ## Read

The Earth spins. This makes it seem as though the Sun moves across the sky during the day.

6 o'clock in the morning   Midday   6 o'clock at night

East   **West**

Put an X where the Sun will be at 9 o'clock in the morning.

Put an X where the Sun will be at 3 o'clock in the afternoon.

 ## What to do

The spin of the Earth makes shadows move during the day.

	Draw the shadow at 9 o'clock in the morning.		Draw the shadow at 3 o'clock in the afternoon.	
6 o'clock in the morning		Midday		6 o'clock at night

 ## Think and do

How can we use a sundial to tell time?

. . . .

Days, months and years

 ## Read

The Earth moves around the Sun. It takes a year.

The Moon rotates around the Earth once every 28 days.

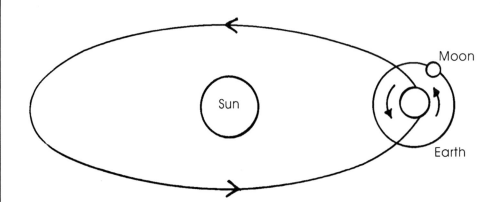

The Earth spins on its axis once a day.

 ## What to do

Answer these questions.

One rotation of the Earth is called a . . . .

How many days are there in a year? . . . .

How many days are there in the month of your birthday? . . . .

 ## Think and do

Which one of these is a **star**?
Which one of these is a **satellite**?

Earth   . . . .

Sun   . . . .

Moon   . . . .

Do any other planets have moons?

. . . .

. . . .

The seasons

Read

The Earth spins once a day on its axis. The axis tilts so that in our summer the North pole is closer to the Sun and the South pole is further away.

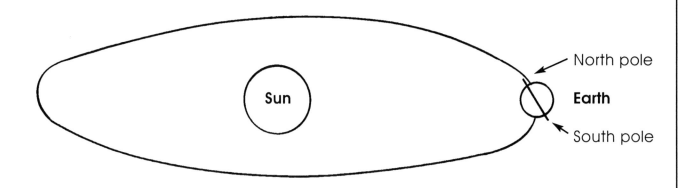

What to do

Make a model to show this (or draw on the diagram above) to show the Earth's position in spring, summer, autumn and winter in Britain.

For example:

Summer in Britain

Winter in Britain

Think and do

Why is it colder in winter?

. . . .

Why is it winter in Australia when it is summer in Britain?

. . . .

. . . .

Space words

What to do

Find the words that we use when we talk about space and the sky.
The words are listed underneath the grid.

s	x	e	a	r	t	h	a	e	g
k	k	s	t	a	r	d	s	m	r
y	o	v	k	v	o	t	t	n	k
o	f	q	s	a	d	w	r	z	i
k	h	g	u	s	a	p	o	o	u
e	e	w	n	m	i	v	n	f	m
t	g	x	a	b	l	c	a	x	o
d	p	l	a	n	e	t	u	k	o
s	p	a	c	e	a	b	t	z	n

sky	astronaut
star	space
sun	planet
moon	earth